How To Read Schematics

A Comprehensive Guide To Reading And Understanding Electronic Circuits

Puma Robin

Table Of Content

Chapter One: Introduction

Definition Of Schematics

Schematics, also known as schematic diagrams, are graphical representations of electronic circuits. They use standardized symbols to represent different components, such as resistors, capacitors, and transistors, and show how these components are interconnected to form a complete circuit. Schematics are used by engineers, technicians, and hobbyists to design, analyze, and troubleshoot electronic circuits, and are an essential tool in the field of electronics.

Why Schematics Are Important

Schematics are important because they provide a graphical representation of electronic circuits and systems, which can be used to design, analyze, and troubleshoot electronic devices. Schematics are essential for electronic engineers, technicians, and hobbyists to understand how circuits work and to identify problems within the circuit. Here are some reasons why schematics are important:

1. **Visual representation:** Schematics provide a visual representation of the circuit components and their connections, which is easier to understand than a textual description.

2. **Design:** Schematics are used during the design phase of electronic devices to plan and visualize the circuit before building it. This can help engineers to

identify potential problems early and make changes before the circuit is built.

3. **Analysis:** Schematics are used to analyze the performance of electronic circuits and identify problems. This can help engineers to optimize the performance of the circuit and improve its reliability.

4. **Troubleshooting:** Schematics are used to identify problems in electronic devices. By analyzing the circuit diagram, engineers and technicians can locate faulty components or connections, and replace or repair them to fix the problem.

Overall, schematics are essential for understanding electronic circuits and devices, and they play a critical role in the design, analysis, and troubleshooting of electronic systems.

The Basics Of Schematic Diagrams

Schematic diagrams are graphical representations of electronic circuits and systems that use standardized symbols and notations to represent components, connections, and their functions. Below are a number of basics of schematic diagrams:

1. **Symbols:** Schematic diagrams use standardized symbols to represent electronic components such as resistors, capacitors, diodes, transistors, and integrated circuits. Each symbol represents a specific electronic function or component.

2. **Connections:** Schematic diagrams show the connections between components with lines that represent wires or conductive paths. The lines are labeled with letters or numbers that indicate the connections between components.

3. **Power and Ground:** Schematic diagrams typically show the power and ground connections for the circuit. The power connection is represented by a positive (+) symbol, while the ground connection is represented by a negative (-) symbol.

4. **Notations:** Schematic diagrams use notations such as voltage and current values, component values, and other information that is important for analyzing and understanding the circuit.

5. **Flow direction:** Schematic diagrams typically show the direction of the flow of current through the circuit with arrows that indicate the flow direction.

Overall, schematic diagrams are an essential tool for understanding electronic circuits and systems, and they use standardized symbols and notations to represent the components, connections, and functions of electronic circuits. Understanding the basics of schematic diagrams is important for anyone working with electronic circuits or systems.

Tools And Materials Needed For Reading Schematics

To read schematics, you'll need the following tools and materials:

1. **Schematic Diagram:** You'll need a copy of the schematic diagram you want to read. This could be in the form of a paper printout, digital PDF, or image file.

2. **Magnifying Glass:** A magnifying glass can be useful for reading small symbols and notations on the schematic.

3. **Pen and Paper:** You may need to take notes or draw diagrams as you read the schematic, so having a pen and paper handy can be helpful.

4. **Multimeter:** A multimeter is a useful tool for testing and measuring components and voltages in a circuit. It can help you verify the values of

components and ensure that they are working properly.

5. **Oscilloscope:** An oscilloscope is a tool that displays the waveform of electronic signals. It can be used to analyze the behavior of signals in a circuit, which can help you troubleshoot problems.

6. **Component Datasheets:** Component datasheets provide detailed information about the specifications and characteristics of electronic components. You may need to refer to datasheets to understand the function and behavior of certain components.

7. **Circuit Simulator:** A circuit simulator is a software tool that can simulate the behavior of electronic circuits. It can be helpful for analyzing and testing circuit designs before building them.

Overall, having a basic set of tools and materials can make it easier to read and understand schematics. Depending on the complexity of the circuit, you may need additional tools or resources to effectively analyze and troubleshoot the circuit.

Chapter Two: Symbols And Components

Basic Symbols And Their Meanings

Here are some basic symbols commonly used in electronic schematic diagrams and their meanings:

1. **Resistor:**

Resistor

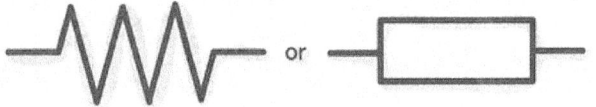

A resistor is a component that resists the flow of electric current. The symbol for it happens to be a zigzag line.

2. **Capacitor:**

Capacitor

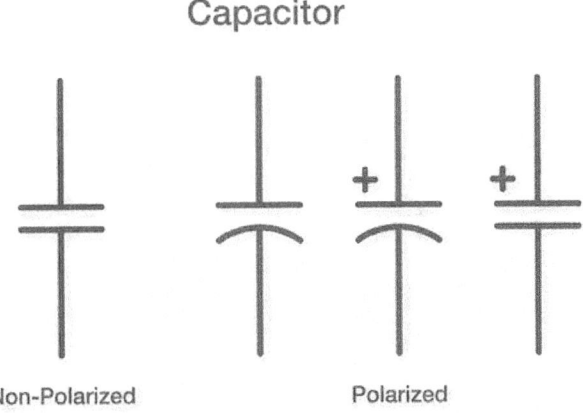

Non-Polarized   Polarized

A capacitor is a component that stores electric charge. It is represented by two parallel lines that are connected by a short line.

3. **Inductor:**

Inductor

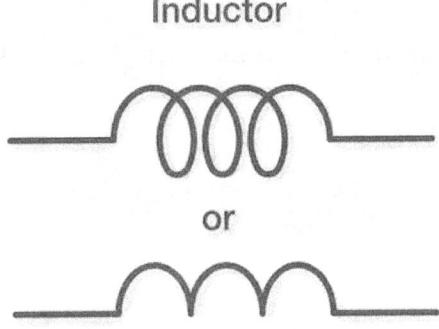

or

An inductor is a component which utilizes a magnetic field to store energy. It is represented by a series of loops or coils.

4. **Diode:**

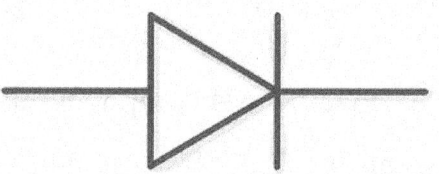

A diode happens to be a component that lets current to flow in just one direction. It is represented by an arrow that points in the direction of current flow.

5. **Transistor:**

Transistor

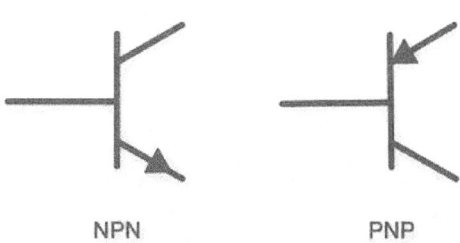

NPN   PNP

A transistor is a component that can amplify or switch electronic signals. It is represented by three regions that are connected by lines.

6. **Integrated Circuit:**

Integrated circuit

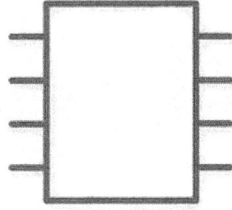

An integrated circuit is a component that

contains many electronic components on a single chip. It is represented by a rectangle with pins or leads coming out of it.

7. **Ground:**

Ground is the reference point for electrical potential in a circuit. It is represented by a horizontal line with three vertical lines that branch off of it.

8. **Power:**

Source

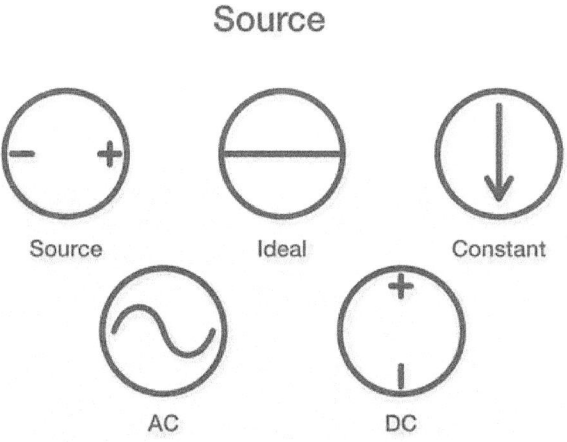

Power is the source of electrical energy in a circuit. In schematic diagrams, power is typically represented by a symbol that consists of a straight line with a circle or triangle at the end. The straight line represents the electrical conductor, while the circle or triangle represents the source of power, such as a battery or power supply.

9. **Battery:**

Battery

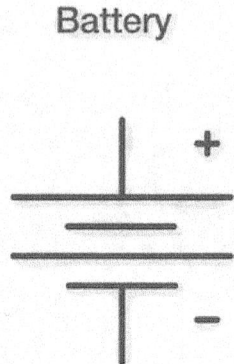

A battery happens to be a source of electrical energy. It is represented by two parallel lines with a plus and minus sign at the ends.

10. **Switch:**

Switches

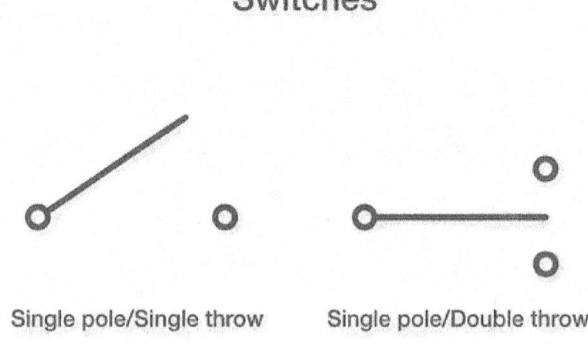

Single pole/Single throw   Single pole/Double throw

A switch is a component that can open

or close a circuit. It is represented by a
line that can be interrupted by a gap or
by a rectangle with a diagonal line.
These are just some of the basic symbols
used in electronic schematic diagrams.
There are many other symbols that represent
different components and functions in
circuits.

Understanding Components And Their Functions

Understanding electronic components and their functions is essential for reading and analyzing schematic diagrams. Here are some common electronic components and their functions:

1. **Resistor:** A resistor is a passive component that resists the flow of electric current. It is used to control the amount of current flowing in a circuit and to reduce voltage levels. Resistors are measured in ohms and their values can be identified by the color code bands.

2. **Capacitor:** A capacitor is a passive component that stores electric charge. It is used to filter noise, block DC current, and smooth out voltage levels. Capacitors are measured in farads or microfarads, and their values can be

identified by the markings on the capacitor or by the code printed on it.

3. **Inductor:** An inductor happens to be a passive component that stores energy in a magnetic field. It is used to filter noise and to smooth out voltage levels. Inductors are measured in henries, and their values can be identified by the markings on the inductor or by the code printed on it.

4. **Diode:** A diode is an active component that allows current to flow in only one direction. It is used to convert AC to DC, to protect circuits from reverse voltage, and to generate radio frequency signals. Diodes can be identified by their part number or by the color code on the diode.

5. **Transistor:** A transistor is an active component that can amplify or switch electronic signals. It is utilized to

amplify signals, to switch circuits on and off, and to control the flow of current in a circuit. Transistors can be identified by their part number or by the markings on the transistor.

6. **Integrated Circuit:** An integrated circuit is a complex component that contains many electronic components on a single chip. It is used to perform complex functions such as signal processing, digital logic, and memory storage. Integrated circuits can be identified by their part number or by the markings on the chip.

7. **Battery:** A battery happens to be a source of electrical energy. It is used to power electronic devices and circuits. Batteries can be identified by their voltage and capacity rating, which is usually printed on the battery.

Understanding the functions of these and other electronic components is essential for analyzing and troubleshooting electronic circuits. By knowing the purpose of each component, you can better understand the behavior of the circuit and diagnose problems when they arise.

Commonly Used Symbols And Components In Schematics

There are many symbols and components used in schematics, but here are some of the most commonly used ones:

1. **Resistors:** A resistor is utilized to limit the flow of current in a circuit. The symbol for it happens a zigzag line.

2. **Capacitors:** A capacitor is used to store energy and smooth out voltage in a circuit. It is represented by two parallel lines with a gap between them.

3. **Inductors:** An inductor is being utilized to store energy in a magnetic field. The symbol for it happens to be a coil of wire.

4. **Diodes:** A diode is used to allow current to flow in one direction only. It is represented by a triangle pointing in the direction of the flow of current.

5. **Transistors:** A transistor is typically utilized to amplify or switch electronic

signals. It is represented by three layers of semiconductor material.

6. **Integrated Circuits:** An integrated circuit (IC) is used to perform complex functions such as signal processing, digital logic, and memory storage. It is represented by a rectangle with pins or leads coming out of it.

7. **Ground:** Ground is used as the reference point for electrical potential in a circuit. It is represented by a horizontal line with three vertical lines that branch off of it.

8. **Battery:** A battery is used to provide electrical energy to a circuit. It is represented by two parallel lines with a plus and minus sign at the ends.

9. **Switches:** A switch is utilized to open or close a circuit. It is represented by a line that can be interrupted by a gap or by a rectangle with a diagonal line.

10. **Transformers:** A transformer is used to
    change the voltage or current of an
    electrical signal. It is represented by two
    coils of wire that are linked by a
    magnetic field.

Understanding these symbols and
components is essential for reading and
interpreting schematics. By identifying and
understanding the function of each
component, you can analyze and
troubleshoot circuits more effectively.

Chapter Three: Understanding The Circuit Diagram

How To Interpret Circuit Diagrams

Interpreting circuit diagrams involves understanding the relationships between the symbols, components, and connections in the schematic. Here are some steps you can follow to interpret a circuit diagram:

1. **Identify the components:** Start by identifying the different components in the circuit, such as resistors, capacitors, diodes, transistors, and so on. Look for symbols that represent these components and try to identify their values or part numbers.

2. **Follow the flow of current:** The arrows in the circuit diagram indicate the direction of the flow of current. Follow the arrows to understand how the current

flows through the circuit and how the components are connected.

3. **Understand the function of each component:** Understanding the function of each component is crucial for interpreting the circuit diagram. For example, resistors are used to limit current, capacitors are used to store energy, and transistors are used to amplify or switch signals.

4. **Analyze the connections:** Look for connections between the components and try to understand how they are related to each other. Are they in series or parallel? Are they connected to a power supply or ground? These connections will help you understand how the circuit works.

5. **Read the labels:** Look for labels on the diagram that provide additional information, such as voltage levels,

frequencies, and values of components. These labels can provide important information about the circuit and help you understand how it works.

6. **Troubleshoot the circuit:** If you encounter a problem with the circuit, use the schematic to troubleshoot the problem. Follow the flow of current and look for any components that might be causing the problem.

Interpreting circuit diagrams can be challenging, especially for complex circuits. However, by following these steps and taking the time to understand the different components and connections, you can gain a deeper understanding of how electronic circuits work

The Different Types Of Circuit Diagrams

There are several types of circuit diagrams used in electronics. Below are a number of the most vital and common types:

1. **Schematic diagram:** A schematic diagram is a graphical representation of an electronic circuit. It uses symbols to represent the different components in the circuit and lines to show how they are connected. Schematic diagrams are often used for design and analysis of circuits.

2. **Wiring diagram:** A wiring diagram shows the physical connections between components in a circuit. It uses pictorial representations of the components and their connections, but does not include the symbols used in a schematic diagram. Wiring diagrams are often used for

installation and repair of electrical systems.

3. **Block diagram:** A block diagram is a simplified representation of a system or process. It shows the major components of the system and how they are connected, but does not include details about the components themselves. Block diagrams are often used for high-level design and analysis of complex systems.

4. **Circuit board layout:** A circuit board layout shows the physical arrangement of components on a printed circuit board (PCB). It includes details about the placement of components, the routing of connections, and the location of holes and other features on the board. Circuit board layouts are often used for manufacturing and assembly of electronic devices.

5. **Piping and instrumentation diagram (P&ID):** A P&ID is a diagram that shows the piping and instrumentation in a process plant. It includes details about the pipes, valves, pumps, and other equipment in the system, as well as the instrumentation used to control and monitor the process. P&IDs are often used in chemical and industrial engineering.

6. **Logic diagram:** A logic diagram shows the logic gates and digital circuits used in a digital system. It uses symbols to represent the different gates and their inputs and outputs, and shows how they are connected. Logic diagrams are often used for design and analysis of digital circuits.

Understanding the different types of circuit diagrams and when to use them is important for effective communication and collaboration in electronics and engineering fields.

Tips For Reading Complex Circuit Diagrams

Reading complex circuit diagrams can be challenging, but here are some tips that can help:

1. **Break it down into smaller parts:** If the circuit diagram is too complex, try to break it down into smaller parts. Focus on one section at a time and try to understand how it works before moving on to the next section.

2. **Identify the key components:** Look for the key components in the circuit, such as power supplies, amplifiers, and filters. Try to understand how these components are connected to each other and how they function.

3. **Follow the signal flow:** Follow the flow of signals through the circuit, starting from the input and moving towards the

output. Try to understand how the signal is processed and modified as it passes through each component.

4. **Use reference materials:** Use reference materials such as data sheets, textbooks, and online resources to understand the functions and specifications of the components used in the circuit. This can help you understand how the components work and how they are connected.

5. **Pay attention to labels and annotations:** Circuit diagrams often include labels and annotations that provide additional information about the components and connections. Pay attention to these labels and use them to help you understand the circuit.

6. **Use simulation software:** If you have access to simulation software, use it to simulate the circuit and observe how it

behaves. This can help you understand the function of the circuit and identify potential issues or problems.

7. Practice and patience: Finally, remember that reading complex circuit diagrams takes practice and patience. Keep practicing and don't be afraid to ask for help or seek out additional resources if you need them.

Chapter Four: Circuit Analysis

How To Analyze A Circuit Using A Schematic Diagram

Analyzing a circuit using a schematic diagram involves several steps. Here is a general approach:

Familiarize yourself with the schematic diagram: Study the schematic diagram and identify the different components, their symbols, and their connections. Make sure you understand the overall function of the circuit.

Identify the power source: Identify the power source for the circuit, such as a battery or power supply. Identify the power source's ratings for voltage and current.

Determine the component values: Identify the values of the components used in the circuit, such as resistors, capacitors, and inductors. Look up the values in reference materials or data sheets.

Apply Kirchhoff's laws: Use Kirchhoff's laws (Kirchhoff's voltage law and Kirchhoff's current law) to analyze the circuit. These laws describe the behavior of circuits and can be used to calculate the voltages and currents at different points in the circuit.

Use Ohm's law: Use Ohm's law to calculate the voltage, current, or resistance of individual components in the circuit.

Use circuit analysis techniques: Use circuit analysis techniques such as nodal analysis, mesh analysis, or superposition to solve complex circuits.

Check your results: Once you have analyzed the circuit, check your results by comparing them with expected values or by simulating the circuit using software.

By following these steps, you can analyze a circuit using a schematic diagram and gain a deeper understanding of its function and behavior.

Using Ohm's Law And Kirchhoff's Laws

Ohm's law and Kirchhoff's laws are fundamental laws of circuit analysis that are used to determine the behavior of electrical circuits. Here's how to use them:

1. **Ohm's Law:** According to Ohm's law, the voltage between two points is precisely proportional to the current flowing through a conductor between them. Mathematically, it can be represented as $I = V/R$, where I is the current, V is the voltage, and R is the resistance of the conductor. To use Ohm's law, you need to know two of the three parameters and can solve for the third. For example, if you know the voltage and resistance of a component, you can use Ohm's law to calculate the current flowing through the component.

2. **Kirchhoff's Laws:** Kirchhoff's laws are used to analyze complex circuits and are based on the conservation of energy and charge. There are two laws: Kirchhoff's Voltage Law (KVL) and Kirchhoff's Current Law (KCL).

● **Kirchhoff's Voltage Law (KVL):** According to KVL, the total of the voltage sources in any closed loop in a circuit equals the total of the voltage drops surrounding that loop. This law is used to calculate the voltages across components in a circuit.

● **Kirchhoff's Current Law (KCL):** KCL states that the sum of the currents entering any node in a circuit is equal to the sum of the currents leaving that node. This law is used to calculate the currents flowing through components in a circuit.

To use Kirchhoff's laws, you need to identify the nodes and loops in the circuit, and then apply the laws to calculate the voltages and currents in the circuit.

Overall, Ohm's law and Kirchhoff's laws are powerful tools for analyzing circuits and can help you determine the behavior of complex circuits.

Calculating Component Values And Determining Their Function

Calculating component values and determining their function is a crucial step in analyzing a circuit. Here are some tips on how to calculate component values and determine their function:

1. Resistor values: Resistor values are typically marked using a color code system. You can use a resistor color code chart to determine the value of a resistor based on its color bands. On the other hand, you can utilize a multimeter to measure the resistance of the resistor.

2. Capacitor values: Capacitor values are typically marked using a code that consists of letters and numbers. The first two digits represent the value of the capacitor in picofarads (pF), while the third digit represents the number of

zeros to be added to the value. For example, a capacitor marked with "104" has a value of 10 x 10^4 pF, or 0.1 µF.

3. Inductor values: Inductor values are typically marked using a code that consists of letters and numbers. The first two digits represent the value of the inductor in microhenries (µH), while the third digit represents the number of zeros to be added to the value.

4. Component functions: To determine the function of a component, you need to understand how it works within the circuit. For example, a resistor is used to limit current flow and can be used as a voltage divider. A capacitor is used to store and release electrical energy, while an inductor is used to store and release magnetic energy.

Overall, calculating component values and determining their function requires a basic understanding of electronic components and their behavior within a circuit. By understanding the role of each component, you can better analyze and troubleshoot circuits.

Chapter Five: Troubleshooting With Schematics

How To Use Schematics To Troubleshoot Circuits

Schematics can be a powerful tool for troubleshooting circuits. Here are some steps you can follow to use schematics to troubleshoot a circuit:

1. **Verify the power supply:** The first step in troubleshooting a circuit is to verify that the power supply is working properly. Check the voltage and current levels at the power source and ensure that they are within the expected range. If there is a problem with the power supply, it may need to be repaired or replaced before proceeding.

2. **Identify the problem:** Once you have verified the power supply, identify the

problem with the circuit. This could be a component that is not working, a connection that is loose, or a broken wire.

3. **Trace the circuit:** Using the schematic, trace the circuit to identify the path that the electrical current takes through the circuit. Follow the path of the current from the power supply to the component that is not working.

4. **Check components:** Use a multimeter to check each component in the circuit for continuity and resistance. Compare the readings to the expected values for each component. If a component is not within the expected range, it may need to be replaced.

5. **Check connections:** Check all connections in the circuit to ensure that they are tight and secure. Loose

connections can cause a circuit to fail or function improperly.

6. **Test the circuit:** Once you have checked all components and connections, test the circuit to see if the problem has been resolved. If the circuit is still not working properly, continue troubleshooting until the problem is identified and resolved.

Overall, using schematics to troubleshoot circuits requires a basic understanding of electronic components and their behavior within a circuit. By following these steps, you can use the schematic to identify and resolve problems with the circuit.

Common Problems And Their Solutions

Here are some common problems that you might encounter when working with circuits, along with their possible solutions:

1. **Short circuits:** A short circuit occurs when two conductive materials come into contact, allowing current to flow through a path with little or no resistance. This can cause excessive current flow, which can damage components and potentially cause a fire. To resolve a short circuit, identify and remove the source of the short, and replace any damaged components.

2. **Open circuits:** An open circuit occurs when there is a break in the circuit, preventing current from flowing. This can be caused by a broken wire or a loose connection. To resolve an open

circuit, identify and repair the break in the circuit, or replace any damaged components.

3. **Overheating components:** Overheating components can be caused by excessive current flow or insufficient cooling. This can damage components and potentially cause a fire. To resolve overheating, reduce the current flow to the component, increase cooling, or replace the component with a higher-rated version.

4. **Incorrect component values:** Using components with incorrect values can cause a circuit to fail or function improperly. To resolve this problem, identify and replace any components with incorrect values.

5. **Incorrect wiring:** Incorrect wiring can cause a circuit to fail or function improperly. To resolve this problem,

check the wiring against the schematic and correct any errors.

6. **Noise and interference:** Noise and interference can be caused by a variety of factors, including electromagnetic interference and power supply noise. To resolve this problem, shield sensitive components, filter the power supply, and ensure that signal paths are well-isolated from noise sources.

Overall, many common problems with circuits can be resolved by identifying and addressing the root cause of the problem. By understanding the behavior of components and following best practices for circuit design and troubleshooting, you can avoid many common problems and quickly resolve any issues that do arise.

Tips For Effective Troubleshooting

Below are a number of vital tips for effective troubleshooting:

1. **Start with the basics:** Verify that the power supply is working properly, check for loose connections, and ensure that components are installed correctly. Many circuit problems are caused by simple errors that can be easily corrected.

2. **Use the schematic:** A schematic can be a powerful tool for troubleshooting circuits. Use it to trace the path of the current through the circuit and identify the components that are not working properly.

3. **Test components:** Use a multimeter to test each component in the circuit for continuity and resistance. Compare the readings to the expected values for each component. If a component is not within

the expected range, it may need to be replaced.

4. **Isolate the problem:** If a circuit contains multiple components, it can be difficult to identify the specific component that is causing the problem. To isolate the problem, remove each component from the circuit one at a time and test it separately.

5. **Keep track of changes:** If you make changes to the circuit during troubleshooting, keep careful notes of what you have done. This can help you track down the source of the problem and avoid introducing new issues.

6. **Take a break:** If you are feeling frustrated or stuck, take a break and come back to the problem later with a fresh perspective. Sometimes a new approach or a fresh set of eyes can help you find a solution more quickly.

7. **Ask for help:** Don't be afraid to ask for help from more experienced colleagues or online forums. Sometimes an outside perspective can help you identify the source of the problem more quickly.

Overall, effective troubleshooting requires a combination of knowledge, patience, and persistence. By following best practices for circuit design and troubleshooting and using the right tools and techniques, you can quickly identify and resolve problems with your circuits.

Chapter Six: Advanced Topics

Advanced Symbols And Components

In addition to the basic symbols and components commonly used in circuit diagrams, there are also more advanced symbols and components that can be used in more complex circuits. Here are a few examples:

1. **Transformers:** Transformers are used to step up or step down voltage levels in a circuit. They are represented by two or more coils of wire with a line connecting them, and a ratio of the number of turns in each coil is typically indicated.

2. **Integrated circuits (ICs):** ICs are complex components that contain multiple transistors, resistors, and other components in a single package. They are represented by a rectangular shape

with pins along the sides, and the specific IC type is typically labeled inside the rectangle.

3. **Diodes:** Diodes are electronic components that enable current to flow in just one direction. They are represented by a triangle shape with an arrow indicating the direction of current flow.

4. **Inductors:** Inductors are components that store energy in a magnetic field. They are represented by a coil of wire with two terminals, and the number of turns in the coil is typically indicated.

5. **Capacitors:** Capacitors are electric field components that retain energy. They are represented by two parallel lines with a space between them, and the capacitance value is typically indicated.

6. **Transistors:** Transistors are components that can be used as switches or

amplifiers. They are represented by various symbols depending on the specific type of transistor.

Overall, these advanced symbols and components can be used to represent a wide variety of complex circuits and systems. By understanding their functions and behavior, you can design and analyze circuits with greater precision and accuracy.

Interpreting Complex Schematics

Interpreting complex schematics can be challenging, but there are several strategies you can use to make the process easier:

1. **Identify subsystems:** Complex schematics often contain multiple subsystems that interact with each other. Start by identifying these subsystems and understanding how they are connected to each other.

2. **Trace the flow of signals:** Trace the flow of signals through the circuit, starting from the inputs and following them through the various components and subsystems. This can help you understand how the circuit is supposed to work and identify potential problems.

3. **Look for patterns:** Look for patterns in the schematic, such as repeated components or subsystems. This can

help you understand how the circuit is designed and how it is supposed to function.

4. **Simplify the schematic:** If the schematic is particularly complex, try simplifying it by breaking it down into smaller sections or using color-coding to highlight different subsystems or signal paths.

5. **Use simulation software:** Simulation software can be a valuable tool for interpreting complex schematics. By entering the schematic into the software, you can simulate the behavior of the circuit and identify potential problems or optimization opportunities.

6. **Refer to documentation:** If the schematic is part of a larger system, refer to the system documentation to understand how the circuit fits into the

overall design and what its specific functions and requirements are.

Overall, interpreting complex schematics requires a combination of technical knowledge, analytical skills, and attention to detail. By using these strategies and taking a systematic approach to the problem, you can effectively analyze and troubleshoot even the most complex circuits.

Designing And Modifying Circuits

Designing and modifying circuits can be a challenging but rewarding process. Here are some steps you can follow to design and modify circuits:

1. **Define the problem:** The first step in designing or modifying a circuit is to clearly define the problem you are trying to solve. This could involve a specific function or performance requirement, or it could be a broader goal such as reducing power consumption or improving reliability.

2. **Choose the components:** Once you have defined the problem, you can start selecting the components you will need for your circuit. Consider factors such as component availability, cost, and performance characteristics such as speed, voltage, and current ratings.

3. **Design the circuit:** With the components chosen, you can start designing the circuit itself. This involves creating a schematic diagram that shows how the components are connected and how the circuit will operate.

4. **Prototype and test:** After the circuit is designed, you should create a physical prototype and test it to ensure that it performs as expected. This may involve using simulation software or actual test equipment to verify the circuit's performance.

5. **Modify and refine:** Based on the results of testing, you may need to modify or refine the circuit design to improve its performance or address any issues that were identified.

6. **Finalize the design:** Once the circuit design is complete and has been thoroughly tested, you can finalize the

design by creating a final schematic and producing a finished product.

Throughout the design and modification process, it is important to keep detailed notes and documentation of each step. This can help you troubleshoot any problems that arise and ensure that the final design meets your requirements. Additionally, it is important to follow safe practices when working with electrical circuits, such as wearing appropriate protective gear and ensuring that all components are properly grounded.

Chapter 7: Schematic Software

Introduction To Schematic Software

Schematic software is a tool that allows users to create and modify electronic schematic diagrams. Schematic software is widely used in the electronics industry for circuit design, documentation, and simulation. There are many different types of schematic software available, ranging from free, open-source tools to commercial products that offer advanced features and support.

Schematic software typically includes a library of electronic symbols and components that can be used to create schematic diagrams. Users can drag and drop these symbols onto a canvas and connect them using virtual wires to create a circuit diagram. Many schematic software tools also offer advanced features such as the ability to simulate circuits, generate bill of materials (BOM) reports, and export schematics in a variety of formats.

Some of the benefits of using schematic software include:

1. Faster design and modification: With schematic software, users can quickly create and modify electronic circuits, allowing them to iterate on designs and prototypes more quickly.

2. Improved accuracy: Schematic software allows users to create highly accurate circuit diagrams, reducing the risk of errors and improving the overall quality of the design.

3. Simulation capabilities: Many schematic software tools offer simulation capabilities that allow users to test and optimize circuits before they are built.

4. Documentation and sharing: Schematic software makes it easy to document and share electronic designs, allowing multiple team members to collaborate on a project and ensuring that everyone is working from the same set of schematics.

Some examples of popular schematic software tools include Eagle PCB Design, Altium Designer, and KiCAD. When selecting a schematic software tool, it is important to consider factors such as the level of complexity required for the project, the cost of the software, and the availability of support and training resources.

Overview Of Popular Software Options

Here is an overview of some popular schematic software options:

1. **Eagle PCB Design:** Eagle is a widely-used schematic and PCB layout software that is popular for its ease of use and affordability. It offers a large library of electronic symbols and components, as well as a user-friendly interface for creating and modifying schematics.

2. **Altium Designer:** Altium Designer is a comprehensive PCB design software that includes advanced schematic capture and simulation capabilities, as well as an extensive library of components and symbols. It is known for its powerful design tools and ease of use, making it a popular choice for professional electronics designers.

3. **KiCAD:** KiCAD is an open-source schematic and PCB design software that

is available for free. It includes a large library of electronic symbols and components, and offers a range of advanced features such as 3D visualization and simulation.

4. **LTSpice:** LTSpice is a free, high-performance simulation tool that is widely used for circuit design and analysis. It allows users to simulate complex circuits and test the performance of electronic designs before they are built.

5. **CircuitMaker:** CircuitMaker is a free, community-driven schematic and PCB design tool that is popular for its intuitive interface and collaborative design features. It offers a large library of electronic components and symbols, as well as a range of advanced features for PCB layout and design.

When selecting a schematic software tool, it is important to consider your specific needs and budget, as well as the level of support and training resources that are available. Many software providers offer free trials or demos, which can be a useful way to evaluate different tools and determine which one is the best fit for your needs.

Tips For Using Schematic Software Effectively

Here are some tips for using schematic software effectively:

1. **Familiarize yourself with the software:** Before you start using schematic software, take some time to learn the basics of the software and its features. Most software providers offer tutorials and online resources to help you get started.

2. **Use keyboard shortcuts:** Keyboard shortcuts can save time and improve your efficiency when using schematic software. Learn the keyboard shortcuts for common actions such as copying, pasting, and zooming in and out.

3. **Customize your workspace**: Many schematic software tools allow you to customize your workspace to fit your

specific needs. Consider rearranging your toolbars and windows to optimize your workflow and improve your productivity.

4. **Use templates and libraries:** Schematic software often includes libraries of electronic symbols and components, as well as templates for common circuit designs. Using these resources can save time and ensure that your schematics are accurate and consistent.

5. **Keep your schematics organized:** As your schematic diagrams become more complex, it can be helpful to organize your components and symbols into logical groups or sections. This can make it easier to navigate your schematics and troubleshoot problems.

6. **Use simulation tools:** Many schematic software tools include simulation capabilities that allow you to test and

optimize your circuits before they are built. Take advantage of these tools to identify potential issues and improve the performance of your designs.

7. **Collaborate with others:** Schematic software often includes collaboration features that allow you to share your designs with others and collaborate on projects in real-time. Use these features to work more efficiently with your team members and ensure that everyone is working from the same set of schematics.

By following these tips, you can use schematic software more effectively and create high-quality schematics that meet your design requirements.

Chapter 8: Conclusion

Review Of Key Concepts

Here is a review of key concepts related to schematic diagrams:

1. Schematic diagrams are graphical representations of electronic circuits that use standardized symbols to represent electronic components and their connections.

2. Schematic diagrams are important for electronics engineers and technicians because they provide a way to understand, design, and troubleshoot electronic circuits.

3. Commonly used symbols in schematic diagrams include resistors, capacitors, inductors, diodes, transistors, and integrated circuits.

4. Understanding component functions is important for interpreting schematic diagrams. For example, resistors are used to control the flow of current, capacitors are used to store and release electrical energy, and diodes are used to allow current to flow in one direction.

5. Ohm's law and Kirchhoff's laws are important tools for analyzing electronic circuits using schematic diagrams.

6. Troubleshooting electronic circuits using schematic diagrams involves identifying and correcting problems such as open circuits, short circuits, and incorrect component values.

7. Schematic software tools such as Eagle PCB Design, Altium Designer, KiCAD, LTSpice, and CircuitMaker can be used to create and modify schematic diagrams, and to simulate and test electronic circuits.

By mastering these key concepts, you can become proficient in interpreting, analyzing, and designing electronic circuits using schematic diagrams.

Tips For Becoming Proficient In Reading Schematics

Here are some tips for becoming proficient in reading schematics:

1. **Study the basics**: Begin by learning the basic symbols and components used in schematic diagrams. Familiarize yourself with common circuit designs and the function of each component.

2. **Practice reading schematics:** Look for schematics of simple circuits and practice interpreting them. Start with circuits that have a few components and work your way up to more complex designs.

3. **Learn the laws and rules:** Learn Ohm's law and Kirchhoff's laws, and understand how they are used to analyze electronic circuits.

4. **Use reference materials:** Keep a reference book or online resource handy for quick reference when interpreting complex schematics.

5. **Collaborate with others:** Join online communities or forums where you can ask questions and learn from others who have experience in reading and interpreting schematics.

6. **Simulate circuits:** Use schematic software tools to simulate circuits and see how they function in a virtual environment.

7. **Practice troubleshooting:** When you encounter problems in circuits, practice using the schematic to troubleshoot and identify the root cause of the issue.

8. **Continuously learn:** Keep yourself updated with the latest developments and advancements in electronics and schematic design. Attend workshops,

seminars, and training sessions to stay current.

By following these tips and continuously practicing, you can become proficient in reading schematics and gain a deeper understanding of electronic circuits.

www.ingramcontent.com/pod-product-compliance
Lightning Source LLC
Chambersburg PA
CBHW071029220526
45467CB00004B/1590